Forgiving Humanity

Also By Peter Russell

The TM Technique

The Brain Book

The Upanishads

The Awakening Earth

The Global Brain

The Creative Manager

The White Hole in Time

Waking Up in Time

The Consciousness Revolution

From Science to God

Letting Go of Nothing

Peter Russell

Forgiving Humanity

How the Most Innovative Species Became the Most Dangerous

The Curse of Exponential Change

Elf Rock Productions

Elf Rock Productions
2375 E Tropicana Ave, 8-733, Las Vegas, NV 89119-8329

Illustrations by Brian Weller

Library of Congress Control Number: 2023931296

ISBN: 978-1-928586-20-3

Contents

... these most brisk and giddy-paced times

~ William Shakespeare

Introduction

All my life I've been fascinated by exponential growth. I first came across it in my math classes at high school and was immediately struck by its implications. The curve of exponential growth not only gets steeper and steeper, it does so forever, leading to mindboggling results. A dollar invested at 10 per cent compound interest per year, would be worth $2.50 after 10 years, $117 after 50 years, and $13,781 after a hundred years. A startling amount. Startling because the human mind finds it difficult to think in terms of exponential change; we are much more accustomed to steady linear change.

I was also keenly interested in the long-term future. Looking at human civilization, I saw that a number of factors were subject to exponential growth—technological development, resource consumption, national economies. If they continued to grow exponentially—and there was every reason to assume they would—we would eventually arrive at situations way beyond what is possible. My teenage mind argued that since "impossible" situations don't occur, humanity would, sooner or later, experience some major disruptions, and probably within my own lifetime. And now, 60 years later, here we are, facing the catastrophic consequences of unbridled exponential growth.

Recent reports suggest we are in the early stages of the sixth major mass extinction in Earth's history—this time caused by one of the planet's own species, rather than by an asteroid impact—and if we don't change our ways radically, and fast, then we, along with many other species, will become extinct in a century or so. *And it is our own fault.*

At least, that is the story we usually hear. Here, I propose a new story of human evolution—not the kind of new story that many people are calling for in which personal and social transformation help us avoid major cataclysms and move on to more sustainable ways of living, ensuring our survival. In this radically different new story, there is no long-term future ahead of us.

We are coming to the natural end of our species' journey, spinning faster and faster into the center of an evolutionary spiral. And, as I will explain, there is no blame for this. It is the inevitable destiny of any intelligent technologically-empowered species.

On the other hand, continued exponential growth means that equivalent amounts of progress will be packed into shorter and shorter intervals. Thus this alternative story does not preclude our achieving as much development in the decades ahead as there has been in the whole of human history so far.

1

How Did We Get Here?

How did the most intelligent and creative species on this planet also become the most dangerous?

On the one hand, we are truly wondrous beings, with extraordinary potential. We have studied the world around us and been awed by what we've discovered. We are aware of our history, of how we came to be here. We can look ahead, imagine a better future, make choices, and reshape the world according to our needs. We have liberated our bodies from much physical toil, freed ourselves from the suffering of many diseases so that we can live longer, more active, lives, and have relieved ourselves of many other burdens. We are capable of love and deep compassion, the appreciation of beauty, the creation of great art, music, and poetry. We find meaning in our lives and have a sense of justice. Nothing like us has ever walked the Earth.

Yet, despite our intelligence, creativity, and technological prowess, we are destroying our planetary support system at an alarming rate. Forests we once inhabited are dying, to be replaced by concrete, wasteland, and desert. The belly of the Earth has been ripped open in an unending quest for raw materials and energy. The air is hazed with pollution. Topsoil is blowing in the wind. Rivers run sour into the sea. While the global climate veers into uncharted territory. In the worst case, the planet will be so changed by our actions that human beings themselves will not be able to survive.

A great tragedy has befallen us. A species with such unprecedented capacities, and the potential to be something truly magnificent, may be about to come to an end.

Where Did We Go Wrong?

Many have tried to identify when we fell from grace. Some see it in the European Enlightenment of the eighteenth century, when human activities took precedence over nature. Others trace it back to the Industrial Revolution, which triggered a burgeoning consumption of natural resources and fossil fuels, with its consequent pollution and eventual global warming. Some blame the colonial oppression of indigenous cultures and the loss of their knowledge and wisdom. Or the legalization of usury—that is, the charging of interest—leading to economies wedded to continual growth. Some see it in the advent of civilization and the movement away from the land to living in cities. Others in the demise of matriarchal societies. Others trace it back to the Agricultural Revolution, when we moved from a hunter-gatherer lifestyle, based on coexistence with nature, to one in which the world was ours to control and exploit. While some argue that the root of the problem goes back even further, to hunting itself. Is it a coincidence, they ask, that many of the large mammals disappeared from the planet around the same time as human beings developed the spear?

Such developments may well underlie some of our present-day woes. But I do not believe any of them are the root cause of our predicament. It's origins go much deeper than any particular human activity or era.

The Seeds of Humankind

The human story began around seven million years ago when *hominin* (the name given to the evolutionary line that eventually led to *homo sapiens*) and *pan* (the line that led to chimpanzees) diverged.

If we were to meet our early hominin ancestors, we'd probably recognize something of ourselves in them— much as we can see something of ourselves in today's great apes: the chimpanzees, bonobos, orangutans, and gorillas. We only have to look at the films of Jane Goodall, who lived with wild chimpanzees in Tanzania, to see many similarities.

All apes are tool users to some degree—as are many other creatures. Chimpanzees use twigs stripped of their leaves to "fish" termites from mounds. They will use rocks as hammers to smash nuts and chop food into smaller pieces, and make sponges from leaves to soak up water to drink. Gorillas will use sticks to test the depth of a stream, and branches to make a bridge across a swamp. Orangutans will make whistles out leaves. If our primate cousins use a variety of tools, it is pretty certain that our own ancestors were doing so way back when they diverged from chimpanzees.

And there are other qualities we share with the great apes. They show empathy for other's feelings, cuddling and comforting each other in times of stress. They play with their young and mourn those who die. They learn from others in their tribe, can solve problems, and have a rudimentary sense of number. They form complex social relationships and possess what is called a "theory of mind," recognizing that others in their group may have their own perception and beliefs. And they pass the mirror test for self-consciousness, being able to recognize the reflection they see as their own.

Yet in other ways we are very different from the great apes. They have not developed the complex scientific, technological, world-changing, global culture that so distinguish us from every other species.

Why, we must then ask, did the hominin lineage diverge so far from that of the chimpanzee? And why so rapidly?

Accelerating Genes

Chimpanzee DNA is 98 per cent the same as human DNA. But within that 2 per cent difference, some genes played a critical role, undergoing greatly accelerated development.

One of them, called HAR1 (Human Accelerated Region 1), had hardly changed over hundreds of millions of years; chickens and chimpanzees have almost identical versions—just two differences. But in hominins it underwent eighteen changes in just five million years—a mere blink in evolutionary time. This gene plays a crucial role in brain development, enhancing the activity of other genes that promote the growth of cells in the neocortex — the structure responsible for cognition and other higher mental functions. Rapid changes in other genes led to the development of the prefrontal cortex, responsible for planning, decision-making, and social bonding.

Another gene that underwent accelerated development led to significant changes in the body, improving our ancestors' physical mobility and dexterity. The foot flattened and the toes shortened, changes that were helpful for walking on two legs. Equally significantly, it led to the fully opposable thumb—a thumb that can touch each of the fingertips.

The hominin hand now had a much better grip. It could grasp objects in different ways and perform delicate operations on them, making it one of the most versatile manipulative organs to have ever emerged. This undoubtedly was a major factor in our ancestors' development of more advanced and innovative tools.

The better grip of the hand also meant they could throw things further and more accurately—probably stones at first, and later spears. No longer did they have to scavenge or catch prey with their own hands; they could fell animals at a distance.

Being able to "reach out" into the world beyond their immediate grasp changed things forever, and in a fundamental way. It sowed the seeds for a growing sense of power over their surroundings. It wasn't wrong; it would have happened sooner or later with any intelligent tool-using species. They were just trying to stay alive. Nevertheless, if there ever was a time when our ancestors first broke from the so-called "natural," or pre-existing, order this could have been it.

Finding our Voice

A quarter of a million years ago *homo sapiens* appeared. And something quite remarkable and unprecedented had happened. It had learned to speak.

Many creatures use language to communicate—the song of a bird, the dance of a bee, electrical signals in fish. But human speech is different. It is a semantic language in which groups of sounds convey specific meanings. The sound "tree" denotes not just an individual tree, but the whole realm of trees, and when incorporated into a grammatical structure can express complex and more nuanced meanings—"the old pear tree at the end of my garden," for example.

As with other major developments in our lineage, speech depended on some significant genetic changes. First, it required the greater processing power of our larger brains, which had been steadily growing since our divergence from chimpanzees. This was further augmented by the FOXP2 gene, often called "the language gene." It, too, underwent an accelerated development, and was responsible for major changes in the brain areas associated with language. It also led to a delicate motor control of the lower face and mouth.

Another genetic change moved the larynx further down the throat, giving the tongue greater freedom of movement than in other primates. This, along with changes in the shape of the mouth, allowed *homo sapiens* to articulate a wide range of sounds, with many subtle variations. Speech as we know it became possible.

All animals learn from experience; but with speech humans could share experiences with each other and so learn, not only from their own experiences, but also from those of their fellows. They could tell each other what they had seen, heard or discovered, and so accumulate a collective body of knowledge, far greater than any individual could. They could also exchange information on making better tools, improving their skills, and performing tasks more efficiently.

Speech also led to richer and more complex social interactions, and a greater sense of community in the tribe. Socialization was an evolutionary advance as significant as tool use itself, and eventually led to humans becoming the most social creatures on the planet, with sophisticated cultures far exceeding that of other creatures. Indeed, it could be said to be one of the hallmarks of being human, and lies at the heart of our collective development.

Speech had ignited what was to become an evolutionary explosion. Its seeds were there in larger brains and the growing use of tools, but on their own these capacities weren't going anywhere fast. Being able to speak and share our discoveries allowed these potentials to flourish, leading to a major acceleration in humankind's speed of development.

Learning to Think

Parallel to these various biological and cultural developments, the inner world of the mind also underwent some major changes—ones equally significant for what was to come.

Speech not only meant humans could talk to each other, they could also internalize speech and have an inner dialogue with themselves—the essence of what we commonly call "thinking." Many other creatures have forms of thinking—being able to anticipate events, solve problems, make decisions, and such—but thinking to oneself in words (and this is the sense in which I will be using "thinking" here) added a new dimension to existing human capacities. They could identify patterns in their experience, form concepts, and entertain new ideas. They could apply reason, draw conclusions, and begin to understand the world in which they found themselves.

They could step back a little from an experience and reflect upon it. They could be aware of feeling sad or happy; of watching a sunset and enjoying the colors; of hearing the sound of a bird and wondering what it might be.

And they could be aware of their own thinking processes. This led to the idea of a self; an "I" that is thinking, that is aware of all this—a self-reflective consciousness that later became the seed for philosophical and spiritual inquiry into who or what we are.

Not only did verbal thinking add new dimensions to human experience, it also expanded our awareness of time—from the immediate present into the past and future.

Memory works largely by association. Seeing a tree, for example, may bring back past images or experiences associated with it. This much probably happens with other animals—a dog may recognize the smell of a tree, which might trigger images of past encounters. But verbal thinking opened entirely new ways of conjuring up memories. With the idea "the tree I climbed as a child," images of the past can be deliberately brought back to awareness. We can relive the story of our lives, and, most significantly, learn from our past and make adjustments in our thinking and behavior.

And we can go back even further, beyond our own personal memory. We can imagine people and events that came before us; go back in time to earlier generations, to our ancestors, and to what came before them. We can even imagine how it might all have begun.

In a similar way, we can think about the future, imagining what we will eat for dinner, where we'll go tomorrow, and what we might do there. We can speculate about what might happen, judge whether or not it might be beneficial for our well-being, consider alternatives and their consequences, and make conscious decisions about what to do.

This brought a freedom of choice unavailable to other creatures, and with it an enhanced capacity for creativity. An entirely new form of innovation had appeared on Earth.

An Innovative Species

Innovation itself was not new. It is the very nature of evolution to innovate, to explore new avenues as they arise. For the preceding seven or so million years of hominin evolution, biology had been evolving, and the principal medium of innovation had been the genes. Now culture was evolving, and the medium of innovation became the human mind — its beliefs, values, skills, ideas and understanding. And these can evolve much faster than genes.

Early humans naturally applied their newfound power of innovation to various challenges they met, in order to make life safer, and better satisfy their needs.

They learned to make sharp edges to stones, creating axes and knives. They shaped points to put on the tips of their spears. Later, bows meant they could aim more accurately and further ahead.

They tamed fire, which not only kept them warm, it also enabled the cooking of food, making nutrients more available and easier to digest. They later used fire to smelt metal, creating better tool-making materials.

Making clothes and building shelters allowed them to better survive inclement weather, and venture into other lands.

Weaving plant fibers into baskets, made it easier to transport food and other goods.

Herding animals decreased the time and energy spent hunting. They maintained grazing lands and guarded their stock.

They began sowing seeds, harvesting the crop, and building stores to preserve food through less abundant times. Furthermore, the cereals they grew provided an additional source of energy.

The invention of the plow made tilling the land much easier. Later, harnessing animals to the plow made it an even more efficient use of human energy, increasing the amount of land that could be cultivated.

As food became more abundant, increasing numbers of people were freed from having to work the land. They could live together in larger communities, divide the labor between them, develop specialist skills, and increase efficiency. Civilization (literally, living in cities) had begun.

Five thousand years ago came the wheel, lightening the burden of heavy loads, and setting the scene for a host of mechanical technology.

During the same period, the fine motor control of the fully opposable thumb led to another world-changing breakthrough—writing.

Limited to speech alone, ideas could not travel far without distortion or loss. Writing enabled the creation of more permanent records. Initially, they were carved in stone, or inscribed on clay tablets. But these were difficult to transport. The development of papyrus, parchment, and then paper, overcame this handicap, allowing their ever-growing body of knowledge to be handed down to future generations, and shared with others in distant lands. The first information technologies had emerged.

Technologically-Empowered Intelligence

Homo sapiens had become a technologically-empowered intelligence. Not only were we using tools, we used them to create more and better tools—the essence of technology.

Creating more effective tools with which to modify and control the world, and using them to get a better understanding of the world, led, in turn, to improved technologies and thence to even more knowledge. Innovation bred further innovation, adding fuel to our evolutionary explosion.

A thousand years ago in China, the first printing press was invented. Manuscripts no longer had to be copied by hand, but could be mass produced, and at a much lower cost. But it failed to catch on the way its re-invention did in Europe, four hundred years later, due to the large number of characters involved.

In Europe, the printing press paved the way for the Renaissance, with significant advances in the arts, philosophy, engineering, and the study of nature, along with an expansion of trade across the world.

This was followed in the seventeenth century by the European Enlightenment. The power of reason gained dominance, leading to the birth of the scientific approach, and the quest for reliable truth. The Earth was no longer flat; nor was it the center of the Universe.

The same period saw the birth of the Industrial Revolution, marrying the power of fire with the utility of the wheel. We today may bemoan some of the repercussions of industrialization, but its founders were a group of visionaries who saw the potential of the steam engine to relieve the load on human muscle, and for clay pipes to improve sanitation and health. Nobody back then knew anything about atmospheric science, or that these technologies, as they came to serve a population exploding far faster than anyone then imagined, combined with their having to serve material desires that back then were inconceivable, would come to threaten the very survival of humankind.

Iron gave way to the much stronger steel, which opened up new possibilities in construction, engineering, and other arenas. Later, synthetic materials were created that pushed the envelope in other directions—stronger, lighter, more fluid, more durable, and more flexible materials. These new materials spawned more new technologies, expanding our power to change the world, and deepening our knowledge and understanding. Which led to more technological advances, more power, and more knowledge.

Information technologies were developing, too. The telegraph made it possible to send messages instantly across the land; the telephone enabled one to speak with someone far away; radio allowed a person to talk to many others; television brought the ability to see others and events across the world.

As technology advanced, so did the power of computation. From mechanical switches, to vacuum tubes and electronic valves, to transistors, to the now ubiquitous microchip; each step pushing speed and storage ahead at an accelerating rate. The now famous Moore's Law showed that over the last half-century computing capacity and speed have doubled every eighteen months or so.

The Internet revolutionized how information and knowledge could be shared. The Worldwide Web took this interconnectivity to everyone. Back then, just thirty years ago, few foresaw that we'd be shopping online, streaming movies, engaging in social media, or any of the other host of online activities that today we take for granted.

Molecular science allowed us to look into the structure of life itself, heralding a revolution in medicine. Examining our own DNA and that of other species opened the door to a much better understanding of our own evolutionary history.

And we could peer into the brain itself—still the most complex information-processing system in the known universe—and begin to understand our own minds.

The rapid progress packed into the last fifty years all came in but one percent of one percent of one percent of one percent, that's one hundred-millionth, of Earth's history!

2

Accelerating Innovation

What does it mean to say the pace of change has been accelerating? Time is not going any faster. The Earth spins around the sun at the same speed. Clocks still tick at the same rate.

It is the rate of innovation—literally, "bringing in the new"—that has been accelerating. We see it today in the rate at which new scientific discoveries are made, new technologies created, new products developed, new social conventions and new skills take hold, and the rate at which existing ideas, technologies and products are continually upgraded or improved—they are all coming faster and faster.

Exponential Growth

This pattern of ever-increasing rates of development is commonly called *exponential growth*. It occurs whenever something's rate of growth is proportional to its current size—a pattern known technically as "positive feedback."

You are probably familiar with audio feedback. This occurs when a microphone connected to an amplifier picks up the sound coming from the speaker, feeds it back to the amplifier, creating an even louder sound in the speaker, which the microphone picks up and feeds back into the amplifier again. As the process repeats, the sound rapidly rises to an ear-piercing shrill.

The human population explosion is another example of positive feedback leading to exponential growth. If birth rates are above replacement level, then the more children that are born, the more parents there will be in the future, and the more children will then be born, and so on. If there are no constraints, this self-reinforcing loop causes the population to grow faster and faster.

Population growth does not follow an exponential curve in the strict mathematical sense, where the rate of growth is a fixed proportion of the current size. Other factors, such as health care and sanitation, also have an impact. Nevertheless, the general principle of positive feedback leading to accelerating growth is still operating. In what follows, I shall use the term "exponential growth" in this looser, more everyday, sense of an "exponential-like growth."

The continuous acceleration in the rate of humanity's development stems from the feedback loop of innovation breeding further innovation. New knowledge may lead to new approaches to handling the world. New technologies can foster further scientific advances, which can lead to more inventions, and other possibilities. Progress feeds back on itself, increasing the pace at every turn.

The Acceleration of Life

The positive feedback of innovation building on previous innovations applies not just to humanity's development; it goes back to the dawn of life.

We don't know just when life on Earth began. The earliest fossils we've found are of simple bacteria from around 4.3 billion years ago. The Earth itself formed some 4.6 or so billion years ago, which means life had likely got started relatively quickly (in cosmic terms).

These simple cells were limited to getting energy either by digesting organic molecules, or from heat in their local environment. Evolution occurred, but much more slowly than in recent eras.

Around 3 billion years ago came photosynthesis, one of nature's more significant innovations, giving life the ability to capture the sun's energy directly.

Around 2 billon years ago, more complex cells appeared with a nucleus encapsulating their DNA. This paved the way for another significant innovation—sexual reproduction. Previously, cells reproduced by splitting in two. Each of the new "sisters" were clones of the original, which meant that genetic changes crept in very slowly. With sexual reproduction, each new cell contained a mix of the genetic information from two parent cells, speeding evolution a thousandfold.

Multicellular organisms, which appeared around 1.5 billion years ago, were another major innovation. Working together in a community, cells could take on specific functions, and form various organs. Evolution was no longer limited to the creation of new types of cells—the muscle cells of fish are not that different from those of human beings. It is primarily the way these organs develop and interact that creates new species of organisms. The reorganization of existing systems can occur faster than the evolution of new types of cell, leading to another speeding up of development. The awe-inspiring diversity of multicellular organisms we see on Earth today came in just the last quarter of Earth's history.

Life moved onto land about 350 million years ago, where it found new evolutionary opportunities. Two hundred-and-thirty million years later, dinosaurs appeared. The first mammals emerged around a hundred-and-thirty million years ago; but most remained small nocturnal creatures, living in trees and burrows, staying out of the dinosaurs' way as much as possible. But when the dinosaur reign ended some 63 million years ago, mammalian evolution took off. And less than ten million years ago, the hominin line appeared.

With the advent of human beings, with their larger cortex, symbolic language, and tool use, the rate of evolution exploded. If the whole of Earth's history were collapsed into a single day, then human beings appeared in just the last few seconds! Civilization a tenth of second ago. The Information Revolution in the last thousandth of a second. And our own lives in an even smaller fraction of second.

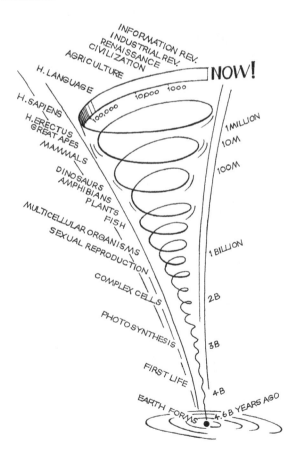

Adding Fuel to the Fire

The fundamental goal of all life, from bacteria to human beings, is to survive, and our newfound power of innovation was, quite naturally, applied to this end. Any development that increased our chances of survival or enhanced our security would have been welcomed and incorporated into the culture of the time. Equally, any invention that allowed us to do more with less effort would have caught on. Indeed, it is hard to think of any significant invention that slowed things down.

An additional feedback loop, this time connecting how we are faring in the world with how we feel inside, has also been increasing the pace of change. If all is OK in our world with no immediate threat or need, we feel OK inside. We are at ease, content. When, on the other hand, there is something we need to take care of—say, for example, we feel hungry or cold—we feel ill-at-ease or discontent. We suffer in some way. This suffering becomes the motivation to attend to whatever is amiss, and so return to a more satisfying state of mind. It feels good to warm ourselves when cold, or eat when we are hungry.

As we got better at managing the world around us and taking care of our physical needs, we had more time to tend to our psychological and social needs, leading us to apply our creativity to increasing our comfort, inner well-being, and personal satisfaction. Innovations that may not have reduced any physical need or threat, but which served more personal agendas or socially conditioned needs also came to be valued and rewarded. In some cultures, the pursuit of happiness itself became the goal, creating a plethora of new commodities and services. All of which further accelerated the pace of change.

Ideally, seeking our own well-being should include the well-being of others, but for various reasons (which I will not go into here), it at times led to an increasing self-centeredness and to decisions that were not always in our best interest. Greed, desire for power, love of money, and other less wholesome traits began to affect the way we applied our creativity and intelligence. Such motivations may, at times, have had significant impacts on the course of human history. But, as we shall see later, such turns would not have changed the overall trajectory of exponential development. Innovation would have continued breeding innovation whatever paths we might have taken along the way.

Today, in addition to these various feedback loops, another factor is increasing the pace of development. We are deliberately trying to speed it up.

Much of humanity has become obsessed with saving time—particularly in those cultures that have experienced the greatest advances in material development. We seek to pack more and more into the hours available. We shop in supermarkets to save the time it would take to visit several stores—and we like quick check-out lines. We build highways around and through cities, or spend fortunes digging tunnels beneath them, in order to reduce journey times, and pack a few more things into our day. And we create ever-faster microchips for ever-faster computers, so that software loads quicker, pages render faster, data is analyzed more rapidly, and a bit more time is saved.

Seduced by the idea of temporal efficiency, we focus much of our creative talent on getting more and more done in less and less time. With the extra time we have "saved," we pack in some more tasks—and then require another time-saving tool to help us cope.

On top of all that, businesses have a competitive advantage if they can develop and market new products or services faster than others. Innovation is itself prized and rewarded—further increasing the rate of change

As a result, humanity has entered a phase of hyper-exponential growth, in which the rate of change accelerates even faster than that of normal exponential growth, further fueling our evolutionary explosion.

An explosion is a self-catalyzing chain reaction. When gunpowder burns, the energy released by just one speck is more than sufficient to ignite the powder around, whose burning releases even more energy, and within a fraction of a second it blows up. The evolutionary explosion of humankind was primed by larger brains and versatile hands, ignited by speech, and spread at an ever-increasing pace across history. It is exploding in all directions—scientific, technological, social, environmental—deep into human culture, and far across the planet.

And, as with any explosion, once started it is unstoppable.

So now the inevitable question demands to be asked: Where is this leading?

3

Our Blind Spot
on the Future

Hundreds of years ago, our ancestors had little concept of progress. Time was measured cyclically—the cycles of days and nights, the moon, the seasons, the years, a lifetime. One generation lived and worked much as the previous generation. There might have been occasional innovations—better food-preserving, sturdier buildings, new ways of hunting—but, generally, the cycles repeated year after year with little change.

With the advent of the Renaissance, the European Enlightenment, and the Industrial Revolution, change came faster. People could remember the days of their childhood, before the printing press, the steam engine, or electricity. Progress had become an intrinsic part of life. We looked back to how things were, and forward to how things could be. Cyclical time had given way to linear time.

Today, progress comes increasingly rapidly. Technological breakthroughs spread through society in years rather than centuries. Calculations that would have taken decades are made in minutes. Communication that used to take months happens in seconds. We look back now, not just to how things have changed, but also to how much faster things are changing. We've entered the era of exponential time.

Our minds, however, find it hard to think in exponential terms; humans evolved in a world where the pace of change, if any, was much slower. As a result, we often fail to see the full implications of exponential change and where it is leading.

You may have heard the story of the king who wanted to reward the inventor of the chess board with anything he wanted. The inventor asked for one grain of rice on the first square of the chessboard. Two grains on the second, four on the third, eight on the fourth, doubling each time till the 64th square would have... how many grains? A mindboggling 18,446,744,073,709,551,615, or about 45 trillion tons of rice. That's a heap as high as Mount Everest—far more than most people intuitively expect.

Money invested at compound interest is another good example of how exponential growth can lead to counter-intuitive outcomes. A dollar invested at 10 per cent per year would be worth $1.1 after one year; $1.21 after two years; $2.59 after ten years; $117 after fifty years; $13,781 after a hundred years; and approximately $2,500,000,000,000,000,000,000,000,000,000,000,000,000,000,000,000 after a thousand years—which is about ten trillion times the weight of the Earth in gold. (This is a primary reason that any economic system based on the charging of interest is destined to eventual collapse.)

In a similar way, we fail to see where exponential rates of change will take humanity. On paper we can perhaps take the acceleration into account; but not in our intuitive ideas of the future. Here, as we shall see, linear time still rules our minds.

A Faster Future

If there is any certainty about the future, it is that the pace of development will continue to grow exponentially. However giddy today's rate of change may seem, tomorrow's world will—barring calamity—be changing even faster. And the world beyond that yet faster still.

We will likely see quantum computers revolutionizing calculation and communication; the study of gravity waves and other astronomical developments bringing a better understanding of the origins of our universe; an integration of gravity and quantum mechanics that could lead to the long-anticipated "theory of everything"; nanotechnology having an increasing impact in industry, chemistry, and medicine; breakthroughs in DNA technologies opening new approaches to health care; the mapping of the human brain becoming as significant as the mapping of the human genome two decades ago, with equally profound consequences for enhancing brain function; artificial intelligence will be ubiquitous; robots, both physical and digital, will become increasingly prevalent; not to mention 3D printed organs and fully-immersive virtual realities.

And there will be unforeseen developments in the application of the Internet. Thirty years ago, few people foresaw the ways we would be using the Internet today, and the rate of innovation now is much faster than it was back then. Some of the coming advances will seem like magic to us today, just as the ability to encode the complete works of Mozart on a speck of matter, and reproduce it in concert-hall quality, would have seemed magic to him.

Approaching a Singularity

Some futurists believe the ever-increasing pace of development will take us into what they call a "singularity"—the term mathematicians give to a point where existing patterns and equations break down and become meaningless.

The North Pole, for example, is a simple geographic singularity: How do you go north from there? Or east or west? And which way is south? Our usual concepts of direction no longer apply.

The center of black hole is another example of a singularity: Space and time cease to have any meaning, and the equations of physics become filled with infinities.

The idea that our own development could be approaching a type of singularity was first put forward both by the mathematician Vernor Vinge, in 1992, and, at the same time, by myself in *The White Hole in Time* (later published as *Waking Up in Time*). More recently the idea has been popularized by Ray Kurzweil, who argues that if computing power keeps doubling every eighteen months or so, as it has done for the last fifty years, then sometime in the next ten years artificial intelligence will surpass the human brain in performance and abilities. These ultra-intelligent systems would then be able to create even more intelligent systems, and do so far faster than any human could.

Kurzweil calls this point in time "the singularity." It is not a true mathematical singularity, in which equations break down or no longer apply; it's what he calls an "historical singularity" — one in which the past patterns of history no longer apply.

As to where that leads, all bets are off. Nonetheless, we can say one thing about a post-singularity world: the pace of change will continue to speed up. Indeed, the emergence of ultra-intelligent systems will accelerate the rate of development even further.

We can't put precise figures to it, but if as much change happens in the next ten years as occurred in the previous twenty (say), then, after the singularity, as much change again might come in the following five. And then as much change yet again in a year or two. Within a short time, the curve would become unbelievably steep, and the rate of change impossibly rapid.

A New Paradigm of the Future

Such unprecedented, and continually increasing, rates of innovation challenge our picture of humankind's long-term future in fundamental ways, demanding a major shift in our paradigm of the future.

A paradigm is a way of seeing the world, and serves as a framework for thinking within a particular field. The Earth-centered model of the universe was a paradigm that lasted for thousands of years, but was replaced by Copernicus's heliocentric (Sun-centered) model in the fifteen hundreds. In the early twentieth century, Newtonian physics gave way to the new paradigms of relativity and quantum physics. In biology, Darwin's theory of evolution through natural selection launched a new paradigm, as did an understanding of continental drift in geography. Not only do paradigms underlie almost every area of thinking, they are seldom questioned. We take their basic assumptions to be the truth.

Paradigms likewise underlie our thinking about the future. Our early forebears saw the future through the lens of cyclical time. They assumed that people would be following similar lives—hunting, farming, building shelter, eating, and socializing much as their ancestors had done—for many generations to come, possibly forever.

When time is seen as linear, and progress is the norm, the future becomes one of steady change. This is the lens through which most people view the future. Some foresee it leading to a transhuman AI reality, others to a world struggling to survive the ravages of climate change; some envision an ecologically sustainable society, others foresee our becoming part of an interstellar community. Yet whatever the scenario, utopian or dystopian, the implicit assumption is that, barring some extinction event, such as a catastrophic change in climate, an asteroid impact, nuclear holocaust, or some other global disaster, the human species will be steadily developing—on this planet or another—for a long time to come. I call this the "future-as-normal" paradigm.

Today, however, we live in exponential time. The rate of innovation will continue accelerating, and by the middle of this century will be far greater than today's dizzying pace. By the next century, it would be unimaginable. The curve would be off the charts. Like the growing mountain of grains of rice on the king's chessboard, it would be way beyond our comprehension—and way beyond any feasible reality.

Yet when we come to imagine our species hundreds or thousands of years in the future, we tend to think in linear time, unconsciously assuming that progress will continue at a fairly steady rate. Take the *Star Trek* scenario, for example, set a couple of centuries from now. The technology on the starship *USS Enterprise* might have seemed futuristic when the TV series first appeared—computers that spoke and understood verbal commands, for example. Today, a mere fifty years later, such features are commonplace. Moreover, the technology does not change much over time. New versions of the *Enterprise* are built, with new capabilities, but the underlying technology is much the same.

But how could that be? Would innovation, the driving force behind accelerating development, have ceased? We have every reason to suppose that science and technology will have continued to progress rapidly. Indeed, given exponential rates of development, the pace of change would have become extremely rapid long before the first *Enterprise* was ever launched. And unimaginably rapid in the years thereafter.

The same is true with just about every other long-term vision of humanity's future. They are not set within the context of exponential change, but are seen through the lens of linear time.

Our Blind Spot on the Future

Someone is said to have a "blind spot" when they are unable to see or consider what is happening; we might have a blind spot in the rear-view mirror of a car restricting our vision, or a blind spot on a person's shortcomings. The term originally referred to a small area of the retina, where the optic nerve enters the eye, that is not sensitive to light. The brain doesn't receive any information about that part of the visual field, so it remains invisible. However, instead of us seeing a blank area, the brain fills in what it thinks should be there, and for most of the time we don't even realize we have this blind spot.

In a similar way, we can have a blind spot on the future. When we think of the future through the lens of linear time—as most of us habitually do—we remain blind to the implications of exponential time. Without realizing it, we make the implicit assumption that the future will be one of relatively steady, linear development. But how could this be, when everything points to the speed of change continuing to grow faster and faster? The two views are inherently incompatible.

Thus, anyone who posits a long-term future for humanity must first give a satisfactory account of how the pattern of accelerating development that has been there since the dawn of our species, and beyond, will now suddenly come to an end.

Since the acceleration is powered by innovation breeding innovation, it would require an end to our inventiveness, creativity, discovery, and problem solving. A most unlikely scenario!

An End to Exponential Growth?

The future-as-normal scenario is so deeply embedded in our thinking that most people either remain blind to the troubling implications of continued acceleration or seek ways to explain how exponential rates of change might come to an end.

A frequent response is that no exponential growth can continue forever. As any particular form of growth approaches its limits, negative feedback comes into play. The rate of growth slows, and the curve flattens out.

A simple example is bacteria growing in a dish. When the numbers are small, there are no effective limits to growth, and the cells multiply exponentially. Then, when they begin to fill the dish, the impending physical limits create negative feedback, which begins to hinder growth. Eventually, when the dish is full, the growth stops.

Or, to give a more familiar example: In the early 1990s, the number of people using mobile phones was low, but nevertheless slowly growing. The Blackberry and flip phones seeded a more widespread adoption. Then, with the advent of the iPhone and Android phones in 2007, the growth curve took off. Ten years later, as mobile phones became more ubiquitous, the number of people still without one decreased and the rate of growth of new users began to slow. Today, as the potential market approaches saturation, the curve is almost flat. (This does not mean overall sales have decreased; the push to upgrade to new models ensures sales continue.)

If we focus on any particular arena of growth, then it is true that most will eventually flatten out. Population, energy consumption, resource extraction, and material development may well reach their respective limits. However—and this is crucial—what we are considering here is not the increasing rate of growth in any particular arena, but the overall acceleration of development. It is this, the pace of innovation itself, that is set to continue gaining speed—not any particular form it takes.

Some propose that future technological advances will lead to more efficient, more elegant and more sustainable ways to satisfy our needs, thus reducing growth in the consumption of resources. Some pin their hopes on advanced AI being able to solve many of our problems, mitigating the effects of climate change, inventing technologies that clean up pollution, or creating economic systems that distribute wealth and resources more equitably. Others argue that a widespread shift in consciousness will lead us to abandon our self-centered materialistic ways.

But even if such changes do come to pass, and we develop truly sustainable technologies and lifestyles, would that bring an end to our accelerating development? No. Innovation would not come to an end. We might choose to apply our creative capacities in different ways, but innovation would still breed further innovation, further fueling exponential change. There can be no returning to the future-as-normal paradigm of steady linear change.

A Maximum Pace of Change?

This doesn't mean there couldn't be some ultimate limit to how rapid change can become. There may, for example, be limits as to how fast the various human, social, and planetary systems can adapt. If so, the pace of change may steady out at some maximum possible rate. Or perhaps the social, economic, and political turbulence produced by increasingly rapid change would eventually limit its speed. However, even if the pace of change were to reach some upper limit it wouldn't mean the pace itself would slow down; only that it would no longer be gaining even more speed.

We could draw an analogy with a car that is steadily accelerating. It will eventually reach a maximum speed—determined primarily by air resistance. The acceleration will then come to an end, but the car doesn't slow down; it continues traveling at top speed.

So, even if the pace of change were eventually to flatten out, we would still be living in a world where change came much faster than today. Imagine change occurring ten times faster (say) than now—and then continuing at that rate for hundreds, perhaps thousands, of years into the future. Hardly a desirable, let alone sustainable, limit.

No Avoiding the Inevitable

Any intelligent tool-using species will seek to improve its lot in life and enhance its safety and survival. Naturally, it would apply its ability to learn, to think about its experience, and to make choices, to its own benefit. Equally naturally, it would seek the knowledge and technologies that allow it to do this more effectively and efficiently. As innovation builds upon innovation, its techno-cultural development will inevitably leap ahead into a phase of hyper-acceleration, with its many unforeseen consequences.

There is no blame for this overall acceleration in our development. We may well blame people, organizations, or cultures for particular turns our path has taken, and for the suffering it has sometimes caused. Different decisions along the way might have led to different outcomes; some possibly less harmful to people, other creatures, or the planet. But the positive feedback loop of new advances facilitating future advances would still have been operating. And, one way of another, we'd have ended up with spiraling rates of growth. This is just the way things are for an intelligent, technologically-empowered species.

There's no reason to suppose this evolutionary pattern, that has been building since the dawn of life, will now slow down. Explosions don't suddenly stop and re-vert to a slow burn.

Thus, rather than holding on to a future-as-normal paradigm based on the implicit assumption of linear time, we must explore how the future looks given the premise of exponential time.

4

The Stress
of Exponential Change

I magine the world twenty-five years from now. Some
things would appear obvious. We will likely have AI
that surpasses the human brain in many of its capabilities. Advances in molecular biology could push life-ex-
pectancy ahead faster than we are ageing. Our bodies,
senses, and mental functions may be augmented in ways
that seem like science fiction today.

Fifty years ahead, the future is much harder to imagine. The technologies at our disposal will be way beyond
anything we can conceive of today. So will the ways in
which we they are used. Who knows what novel capac-

ities will emerge? In addition, the pace of change will have ramped up even more, taking us ever faster into completely uncharted territory—a world that is as unimaginable to us today as our world would have been to the ancient Greeks.

Danger in the Wings

Given the unprecedented opportunities in such burgeoning development, it is easy to see the future in rosy terms. Yet, at the same time as we forge ahead into ever greater technological prowess, some of the repercussions of continued growth are also becoming apparent.

Take the human population explosion, for example. Its seeds were sown two million years ago, with the shift to a hunter-gatherer lifestyle and a better diet. People lived longer and more children survived. Slowly, at first, the population began to grow. At the beginning of the Agricultural Revolution, it had reached a few million. Ten thousand years later, as the first civilizations emerged, it was approaching two hundred million. At the start of the Industrial Revolution, seven hundred million. In 1900, one-and-half billion. Today, we stand at over eight billion, presenting challenges for food, water, housing, and geopolitics that were unforeseen a few centuries ago.

Growth in other areas has led to further challenges, with the result that we've had to apply our intelligence and creativity not just to meeting needs and improving well-being, but also to solving the various problems that have arisen along the way. But the solutions we came up with often created graver problems. Solving these required us to find ever more sophisticated solutions, with their own unforeseen challenges.

Farming the land, for example, exploited a biological fertility that had taken eons to accumulate. The more intensive agriculture became, the more fertilizer needed to be added, and the more the soil deteriorated. Today it has reached the stage where some soils have become so depleted little can be grown.

Using wood for fire and construction seemed an obvious and innocuous step. But deforestation reduced once verdant areas to desert, and radically changed the ecology in others.

Burgeoning empires solved some problems of supply by harnessing resources from across the world—which led to new social and political problems.

Energy constraints were solved by burning fossil fuels. The unforeseen problems that this "solution" created are now plain for all to see. Let's hope we don't try to address the challenge of climate change by tinkering with the Earth's atmosphere or oceans. The ramifications could be disastrous.

Focused as we were, quite naturally, on our current needs, we could not see the long-term impact of our problem-solving. This was not just a lack of foresight; we simply did not know what the impact would be; the data didn't exist. Who at the dawn of the Industrial Revolution could have imagined that burning fossil fuels would pose a threat to our continued survival? The first hint of the dangers did not come until the late nineteenth century, and even then were almost completely ignored.

The Stress of Acceleration

And there was an even greater danger waiting in the wings—one that is often overlooked—namely, the increasing stress that accelerating change places on ourselves and the world.

Stress in general may be defined as a failure to adapt to change. In human terms, the more we have to attend to, plan for, worry about, and take care of—that is, the more to which we have to adapt—the more likely we are to suffer stress, with its various effects on our physical, mental, and emotional health, not to mention repercussions on family, friends, and colleagues.

Today's increasing pace of life is itself a growing source of stress. There's more information to keep abreast of, more technologies to master, more challenges to meet, more skills to learn, more tasks to accomplish. Some of us find ourselves having to work longer hours, some at weekends, too. We may feel overwhelmed, under increasing pressure to make quick decisions, have more and more things to do—and less time to do them all in. The amount of quality time we can have with ourselves, family, and friends, relaxing and recovering from such pressures, is getting less—and for some disappearing completely.

Moreover, it is not just people who are experiencing the stress. Our social, economic, political, and environmental systems are all being impacted as they fail to adapt to ever-faster rates of change.

A Crisis of Acceleration

When we look at the many crises facing humankind, it is easy to blame various apparent causes—over-consumption, excessive waste, financial greed, poor government, unsustainable policies, etc. These are all real issues, each with serious impacts on humanity and the planet. But behind each, in one way or another, is a deeper issue: the ever-increasing pace of change.

Oil reserves are running low because we are consuming them a million times faster than they were laid down. Similarly, mineral resources such as platinum, copper, zinc, nickel, and phosphorus—all of which are crucial for contemporary technology—will become seriously limited within a few decades. Yet our demand for physical resources continues to grow exponentially, exacerbated by the rapidly growing needs of developing countries.

This depletion of resources will not only impact humanity's future; it will also be felt far into the planet's future. Most mineral ores were laid down early in Earth's history by microorganisms that concentrated particular elements in their cell structures. Should some new intelligent species emerge after humanity has gone, it will not have the luxury we have had of many easily-accessible minerals. We have benefited from a once-in-a-planet's-lifetime opportunity.

On the other side of the equation, rapid growth in industrialization has led to an accelerating profusion of pollutants in the sea, air, and soil. Toxic rivers kill much of the life both within them and where they enter the sea. Plastic accumulating in the oceans is entering the food chain. Noxious substances circulate in the atmosphere, with untold consequences for the living systems that ingest them. Not to mention the buildup of that seemingly harmless gas, carbon dioxide.

The ever-increasing concentration of carbon dioxide in the atmosphere results from our accelerating consumption of fossil fuels. Previously, plants and oceans absorbed most of it, but we are now releasing the gas hundreds of times faster than these systems can handle, leading to a rapid rise in the Earth's temperature, with serious repercussions in many areas, from food and habitation to commerce and geopolitics. We might, if we really put our hearts and minds to it, avert the most damaging repercussions of climate change; but climate change is just one of the critical challenges now facing humanity.

I've already mentioned the inherent instability of any economic system based on compound interest. Such systems need a continual growth in net wealth, partly to avoid inflation, and partly to repay the interest on all the money out on loan. A three-percent annual growth in GDP may be deemed healthy for a nation's economy, but the long-term impact of such growth is devastating. Compounded over a hundred years, three-percent annual growth would lead to a consumption of energy and resources, and consequent pollution, at *twenty times* today's rate! Unsustainable, to say the least.

In addition, some of the current geopolitical instabilities can be traced back to accelerating change. When Europeans began colonizing other parts of the world, they took their more advanced scientific, technological, social, and political systems to peoples whose development was a thousand or more years behind—and growing much more slowly. The dangerous consequences of this are now apparent in regions of the world where people still live with customs and values of pre-industrial Europe, yet have access to modern weaponry, internet, and ease of travel. We are experiencing not so much a clash of cultures, but a clash of eras—a clash originating in a mismatch in their rates of progress.

A system can tolerate only so much stress before it breaks down. If a wheel is made to spin faster and faster, the increasing stress will eventually lead it to fly apart. In a similar way, as the pace of change grows ever faster, the systems involved—whether they be our own biological system, social, economic, and political systems, or the planetary ecosystem—will eventually crack and break-down. Crises will pile up on each other faster and faster, heading us into the perfect global storm.

Spiraling into the Eye of the Storm

We could liken our situation to water swirling towards the plughole in a basin. Far from the center, the water is moving slowly, almost imperceptibly, perhaps taking a minute to complete a revolution. Halfway to the center, it is moving four times as fast. Halve that distance and it is moving four times as fast again, a revolution every four seconds. Halve that, and it's whirling around once a second.

Humanity is whirling faster and faster on its own evolutionary spiral. And, just as the ever-more rapid whirling of the water comes to an end when it reaches the center of its vortex, our accelerating pace of development will finally come to its own end. But it will not end because we change our ways or get innovation under control. It will come to an end as we spiral into the center of our temporal whirlpool—into the evolutionary singularity we inevitably started heading toward as soon as the power of innovation was put into our hands.

We thus arrive at what is, initially, a most uncomfortable conclusion. When we consider the future from the perspective of exponential rather than linear change, it appears that technologically-empowered, innovative societies are inevitably short-lived.

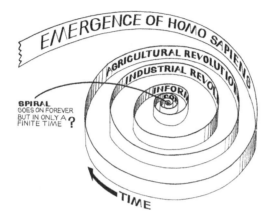

They are short-lived, not because of any fault in their people, nor in their use of technology—but from the acceleration itself. Innovation will keep breeding innovation, furthering not only exponential growth, but also exponentially increasing stress on the biological, social, and planetary infrastructures. Eventually, they can no longer hold, and collapse.

Across the Universe

The physicist Enrico Fermi pondered the apparent contradiction between a high probability of extraterrestrial civilizations existing elsewhere in our galaxy, and the lack of evidence for, or contact with, such civilizations. Why haven't they already colonized Earth? Or why don't we detect their radio transmissions?

Many answers have been proposed, ranging from the possibility that they are already here, to the distances being so vast that the journey would take millions of years. But the true answer may be that they don't exist. Or to be more precise, they do exist, but for only a brief flash in cosmic time.

When we ponder the possibility of extraterrestrial civilizations, we usually make the mistake of thinking in terms of linear time. We tacitly assume they would continue for thousands, or even millions, of years in a relatively stable state. They might make advances from time to time, but not at the rapid rate we know today—let alone the even more rapid rates of tomorrow. We don't consider that their own development will likely be on a similar exponential curve to ours.

Whatever their physical form, any intelligent tool-using species is likely to develop the knowledge and technologies that enhance their safety and survival. The more they learned, the better their tools, the smarter they became, the faster they would develop. As innovation built upon innovation, their rate of progress would keep gaining speed. Thus any species more advanced than us would likely experience change far faster than we do now. Within a short time (evolutionarily speaking), they would meet the consequences of hyper-acceleration, spiraling into the center of their own evolutionary whirlpool.

On the other hand, there may well be advanced intelligence out there that has not taken the technological path. Here on Earth, whales and dolphins show signs of intelligence approaching that of humans — even surpassing us in some respects. They've been around for tens of millions of years — long before our primate ancestors came on the scene. But not having hands, they have not created tools and technology, so have not got caught up in a spiral of ever-accelerating development.

Perhaps the evolution of intelligence on some other planets has taken a similar, non-technological, course. Advanced intelligence might be living in a planet's oceans (whether they be oceans of water, methane, or some other liquid). There, suspended weightless, a creature's body would be free from the constraints of gravity, and could grow much larger than on land, opening the possibility for even larger brains. It might be there, in extraterrestrial oceans, that intelligence and awareness far surpassing our own has evolved.

Doom and Light

Our future in exponential time is not, however, all doom and gloom. The conclusion that humanity's evolutionary explosion is destined to end — and in the not-too-distant future — may, at first sight, seem to imply an end to the many scientific and technological advances on the horizon, and an end to all that we hoped we'd become.

Seen through the lens of linear time, this would indeed seem to be the case. At today's rate of progress, we might well think it would take centuries, or perhaps millennia, for our species to achieve all we imagine possible. From this perspective, the continued advancement of our species demands we mend our ways, live more sustainably, and take better care of the world. Otherwise, if the global ecosystem collapses, that vision of a long-term future will expire and we will never achieve our full potential. Through the lens of linear time, it's a race between breakthrough and breakdown.

However, from the perspective of exponential time—which is the lens through which we must now consider the future—things look very different. With accelerating development, the time between significant advances will be compressed into shorter and shorter intervals. Progress similar to that of the last fifty years may in the future take only twenty years, and a similar amount again may take only ten years or so. We will see technological advances way beyond those which we now imagine, plus equivalent advances in scientific understanding, all packed into ever shorter periods. Who knows what we might then discover or create? Almost anything is possible.

Breakthrough and breakdown are now two sides of the same coin. They are ramping up together and coming to a head together. No longer is it a question of "either-or," but an acceptance of "both-and." In the coming decades, we will see technology beyond our dreams, in a world that's breaking at the seams.

We started this exploration with the question of how is it that the such an intelligent and innovative species has become the most dangerous? The answer, it is now becoming clear, is the two go hand-in-hand.

The acceleration in our evolution that came from our enhanced intelligence and creativity, inevitably leaves in its wake an increasing turbulence that now threatens to bring the system down. This doesn't mean we somehow went wrong; this is how it is to be a technologically-empowered intelligence spinning ever-faster into the eye of its evolutionary hurricane.

5

Facing Reality

Not surprisingly, most people have great difficulty accepting that technologically-empowered intelligence is intrinsically short-lived. It's the last thing we want to hear. We know that our species will not last forever, but most of us have imagined the eventual end to be way in the future—many thousands or millions of years from now. We think this intelligent, creative, self-aware being ought to be around for the long-term. So it can be extremely challenging to realize that our collective end may arrive much sooner than expected.

It is hard to find words that adequately capture the immensity of where we are headed. We are facing the end of a socio-cultural, scientific, technological development that has been going on since the appearance of homo sapiens, and whose roots can be traced back even further. Humankind has never faced anything like this before. History shows that civilizations come and go; each failing in its own time. But there is no precedent for the ending of a global civilization, and possibly that of our own species too.

And, perhaps even harder, to accept that this is inevitable. The seemingly benign process of innovation breeding further innovation has led inexorably to unsustainable rates of change, and thence to the breakdown of the various systems essential to humanity's continued existence. And with no feasible way to avoid it.

Facing Death

Obvious parallels exist with our own death. We know it is coming, but unless we are diagnosed with some terminal illness or suffer a potentially mortal injury, we tend to push it away to sometime in the future—not tomorrow or next week. As a result, many of us live in a state of denial. We know death is inevitable; yet seldom face its reality, often living as if it will never happen.

Accepting our personal mortality is, however, part of being a mature human being. Indeed, confronting death directly can produce profound shifts. People may reconsider what is important, value love more than wealth and fame, and find a renewed purpose and greater appreciation of life.

Here we are facing the end, not just of our personal lives, but also the potential end of our species. When we look at all that we have created, all the good there is in us, all that we hold dear, and all that we might yet become, it is hard to accept our species may be approaching its own inevitable fate.

To make matters worse, what little future there may be does not look as rosy as we might have hoped. The increasing strain of exponential growth on various human, social, and ecological systems point to things coming to a head this century. Or rather, I should say "increasingly coming to a head," since the impact of this stress is already apparent in today's world. Hardly welcome news for younger generations who, even now, view the future with growing despondency. Or for parents, as they picture their children and grandchildren growing up in a world very different from the one they'd imagined they would inherit.

The Great Unraveling

We, today, are witnessing the beginning of the great unraveling.

Scientists express growing concern that climate change may have passed a tipping point. Even if we were to stop all fossil-fuel burning today, global temperatures would continue to rise for decades, probably triggering a runaway greenhouse effect as the much more potent greenhouse gas methane is released from the tundra and the deep ocean. The warmer the planet gets, the more methane is released, and the more the planet warms— the familiar positive feedback loop underlying exponential change. In addition, the more ice that melts, the less heat is reflected away from the planet and the more is absorbed by the newly uncovered land and ocean areas, creating another positive feedback loop.

The impact on the world's ecosystems will be profound, which in turn will have a major impact on humanity. As drought and heat turn large areas of arable land into desert, there will be unprecedented crop failures and famine leading to mass migrations as millions are forced to abandon their homelands, creating huge challenges for the regions to which they are fleeing. Increasingly severe storms and their aftermath will take a growing toll. Extreme heat waves in regions with little water or air-conditioning will be catastrophic. Devastating wild fires will ravage more and more land.

Moreover, climate change is but one of many intertwined crises. Economic collapse, growing poverty, and unprecedented natural disasters, will likely lead to widespread social breakdown and the rise of authoritarian regimes. Global conflicts will increase as food, water, and other resources become increasingly scarce. Epidemics of viruses and drug-resistant bacteria, biological and chemical terrorism, collapse of the Internet through hacking or cyber-war, systemic chaos... all are possible. Doubtless many will happen.

And, more than likely, completely unforeseen events will take their toll.

Collective Grieving

As the reality of the unraveling hits home, there will be widespread despair, depression and distress. We can expect increasing anxiety and anguish over how our own lives might unfold; and grieving over what has become of us, this wondrous, creative, intelligent species, and of this amazing planet with its awe-inspiring beauty and diversity of life.

How will we respond to the heartbreak we'll feel as we head into the eye of the storm? Even now, it is hard to fully let in the increasing tragedies around world, and the suffering that more and more people are experiencing, not to mention that of the numerous other species caught up in our destiny. How will we process such unprecedented grief? Will we remain in denial, refusing to accept what is unfolding? Party wildly, consuming to the last drop of oil? Bury our heads in horror and hopelessness? Or find the acceptance that allows us to move into the unknown with courage and an open heart?

With the death of a loved one, people tend to pass through several stages to grief. The first is denial. We can't believe he or she is no longer with us. No. It can't be true. Then comes anger. Whether directed towards God, a circumstance, a person, an illness, or some other agency. How dare this happen? It's not what I expected or wanted. Third can come bargaining. We want our loved one restored. If only I had just done this or that. Maybe there's a way to bring them back. This is often followed by depression. We withdraw from life, consumed by sadness, wondering if there is any point in going on alone. Finally comes acceptance—accepting the reality that our loved one is gone. We may not like this new reality, but we adjust and learn to live with it.

As the direness of our predicament becomes increasingly apparent, humanity will experience its own collective grieving. The loved one now is not another person, but humankind itself.

Denial is already apparent—"denial" here meaning more than a refusal to accept something may be true, but also choosing to look away from what is clearly happening and not consider where it might be headed. There is denial of the impact of climate change, denial of the poverty that afflicts one third of us, denial of the fragility of civilization, denial of the immense suffering in the world. And now, potentially, denial of the consequences of exponential growth.

Those who come out of denial may move into anger: anger at the corporations, the politicians, the wealthy, the Church, the military, the terrorists, or anyone else we blame for the crises. They have messed things up, spoiled our future. How could they have been so blind?

Already we see signs of the third phase, bargaining. If we just change our ways, perhaps we could make things okay again, rescue ourselves from the tragedies that lie ahead. Perhaps it is not too late to clean up our act and save the world.

Then will surely come depression and a deep sadness at what has befallen us. What have we done? This is terrible. The future looks so bleak. Even now, psychotherapists report that, in addition to a client's personal issues, there is a growing angst about the state of the world and where we are headed.

Finally—hopefully—there will come acceptance of our predicament. We let go of our attachment to how things should be, our hope that things will turn out well in the end. We don't deny the painful emotions that arise, but accept them as part of living through these times. Accept this is how it is for a technologically-empowered intelligence, as it spirals towards its inevitable destiny.

A Sixth Mass Extinction

There have been five major mass extinctions, and several minor ones, in Earth's history. Nearly every species becomes extinct eventually; but in a mass extinction a large percentage of the species become extinct at the same time. Two hundred and fifty million years ago, about 90 per cent of Earth's species died out in a period of rampant volcanism. But life is resilient; it adjusted and moved on. In the most recent mass extinction, about 65 million years ago, nearly 80 per cent of the species died out, including most of the dinosaurs, when an asteroid struck the planet. Tragic as it might have been at the time, it created an evolutionary opportunity for mammals, leading eventually to the great apes, to hominins, and to us today.

Now we are in the beginning of a sixth mass extinction, one triggered this time not by an asteroid or volcanoes, but by one of the Earth's own creatures—the human being.

It is estimated that up to a hundred and fifty of Earth's ten million species are currently lost each day. That would amount to 10 percent of all species each decade—which is many thousands of times faster than the natural extinction rate. Moreover, the number of species threatened with future extinction is steadily growing. Forty percent of amphibians, twenty-five percent of mammals, thirteen percent of birds, forty percent of insects, and forty percent of the world's plant species are now thought to be at risk. And as ecosystems deteriorate further, the numbers threatened may well increase.

How catastrophic will it become? Who knows? But there is no doubting a sixth mass extinction is now underway.

Mass extinctions don't, however, happen overnight. A colliding asteroid may instantly destroy life in the area it hits, but the ensuing planetwide die-off occurs over many years. Species become extinct as their habitat becomes increasingly inhospitable; their numbers begin to decline, until eventually only a few remain, and then none. Similarly with humankind. It is unlikely the human race is going to be suddenly wiped out in some planetary cataclysm—the breakdown of ecological systems is much more likely than an asteroid impact. As things heat up, some of us might die prematurely, and possibly from unanticipated causes. Others might still live to a ripe old age; but in a world that is very different from today's—one with, among other things, a now *declining* population.

Some of us may possibly survive, perhaps eking out an existence in the newly-green polar regions, or possibly in some contemporary arks—self-sufficient, sustainable, high-tech habitats created by the wealthy to ensure their survival in the final days. If they are lucky, they might even survive long-term.

But we would still be an innovative species. We would still be seeking ways to improve our lot—which in such a future would not be a very happy lot. As before, we would find ways to survive better and more comfortably. The positive feedback of innovation breeding innovation would still be operating. Slowly but surely, the spiral of acceleration would begin to wind itself up again, and, slowly but surely, we'd eventually approach a similar point in time.

Even if some indigenous people survived, the ultimate fate would probably be the same. It is true that indigenous peoples today generally live in harmony with their environment, and with a wisdom we often admire. But remember that we in the developed world are the descendants of people who, thousands of years ago, were "indigenous" to their own lands, and likely had values and lifestyles similar to those we see in indigenous peoples today. But over time, as we "progressed" into the modern world, most of that wisdom was lost and replaced by the more materialist culture we live in today. The same could eventually happen to today's indigenous peoples as they developed better ways to stay safe and survive. A hundred thousand years from now they could be facing the consequences of their own exponential growth.

The Light Still Shines

Such scenarios may seem devastating, and without any trace of hope; yet, at the same time as our social and ecological systems are breaking down, science and technology will continue making major breakthroughs. We could see as much, or even more, development packed into the decades remaining as there has been in the whole of human history so far.

We touched on some likely technological advances earlier: Quantum computers revolutionizing calculation and communication. Nanotechnology having a major impact in industry, chemistry and medicine. Developments in genetic technologies opening new approaches to health care. Replacement body parts grown from stem cells. Our bodies, senses, and mental functions being augmented in ways that seem like science fiction today. The continued evolution of the Internet bringing capabilities as far beyond today's streaming movies, social media, and multiplayer games as these were beyond the Internet of twenty years ago. Artificial General Intelligence (AGI) will likely be attained—that is an artificial intelligence that can accomplish any mental task that a human being can, and probably better.

Some of these new technologies may help make life more comfortable and efficient, improve health care, manage the environment better, reduce material consumption, and soften some of the impacts of the unravelling. But they are unlikely to reverse humanity's trajectory; more likely the opposite as innovations continue to inspire more innovation, winding up the spiral of development even tighter.

Scientific knowledge will likewise continue to expand. Physics is approaching a "theory of everything"—a set of mathematical equations that underlie all the forces of nature. We are not there yet. But many think we might be in a few decades, or perhaps less. In cosmology, we are beginning to understand how the universe in which we find ourselves came into being, and where it might be heading. Again, we are not there yet. Many unanswered questions remain, and almost as many competing theories. But it is conceivable that we may answer these questions before too long.

The area where we still have the most to discover is our own brains. In the coming years, the mapping of the human brain will become as far-reaching as the mapping of the human genome two decades ago. It will spawn not only a much deeper understanding of brain function and new approaches to mental illness, but also technologies that enhance our mental abilities. It could, moreover, offer insights into the relationship between brain and consciousness—a realm where we still know very little, if anything.

Just as science and technology have liberated us from various physical limitations, so, too, the exploration of our inner world could liberate us from psychological handicaps. For more than a century, psychologists have shown how a person's past can lead to behavior that is not in one's own best interests—nor that of others. Social conditioning, childhood trauma, unmet needs, and other factors can lead to psychological issues that impede our lives and our relationships. These "scientists of the mind" have developed various processes and approaches—one might call them "mental technologies"—that can alleviate these issues and help people lead more caring and fulfilling lives.

Conventional approaches, however, can only go so far. Consciousness is primarily a subjective experience and thus can, and should, also be explored directly, from within. Various mystics, philosophers, and sages have trod this introspective path, exploring the nature of awareness itself. Time and again they've affirmed that beneath our day-to-day experience lies a deeper sense of being, unperturbed by the goings on in the world, and our hopes and fears about them. They have developed meditative and self-enquiry practices that lead to greater calm and self-awareness, and claim that the more in touch we are with our inner being, the more considerate, compassionate, and caring we become—qualities that could prove invaluable in meeting the challenges ahead.

Interest in these subjects has been growing steadily. Self-help and personal development books dominate the market, while ever more videos and courses are appearing online. Interest in meditation and mindfulness is exploding, with teachers of a thousand flavors springing up everywhere.

The more that people explore ways to liberate the mind, and the more they learn, the more they can help and inspire others. Another example of positive feedback leading, once again, to exponential growth—this time, exponential growth in knowledge of our own minds and the nature of consciousness itself.

I do not believe this movement will change our overall trajectory (as some of its more ardent exponents hope) leading us to mend our ways and live in harmony with the planet. It is not going to take us off the ever-tightening spiral of change. It could, however, lead to some major shifts in consciousness that help us navigate the times to come with less attachment to outcomes, more inner flexibility and stability, and greater care and compassion.

An Exponential Reminder

You may not agree with every aspect of the scenarios I've laid out; you might well see possibilities that I have not. But whatever your sense of how things could unfold, I urge— nay, insist—you look from the perspective of exponential change. Our thinking is so embedded in linear change, it is easy to fall back into "future-as-normal" scenarios, with their implicit assumption of steady development.

This blind spot on the future means we must be constantly vigilant that however we picture our future we do so in the context of spiraling rates of change. There is no escaping this. It has been building for millions of years. And is not about to stop.

6

A Blossoming
of Consciousness

What will happen as we spiral faster and faster into the center of our evolutionary whirlpool? The overall trend may be fairly clear, but how events will actually unfold is far from certain. Moreover, as changes are compressed into shorter and shorter intervals, the future will become ever more difficult to predict.

Rather than trying to determine what will or will not happen, we would be better preparing for a future that is becoming increasingly uncertain? With events we can predict, as, for example, an impending hurricane, we have some idea of what will happen and how to prepare. But how do we prepare for the unpredictable?

One thing that will help is greater resilience—usually defined as the ability to withstand or recover from setbacks. In this case, the ability to withstand or recover from the disruptions and challenges of unanticipated changes.

As far as physical resilience is concerned, there's a variety of things we can do. I won't go into them here since the options depend very much on our personal needs and circumstances. But one general approach that can be helpful is to ask yourself: If a particular scenario happened, what is one thing I'd wish I had done? And then, if possible, do it. Later, you could drill deeper, or ask a similar question about another scenario.

We can also cultivate greater mental resilience. Trees in a storm provide a good analogy. If a tree is to withstand the storm it must be flexible, able to bend with the winds; a rigid tree will soon blow down. And it must have strong roots, be stably anchored in the ground.

The same is true for us. If we are to survive the coming storm of change—along with some exceptional gusts—we will need to be more flexible. We'll need to let go of our attachments to how things should be; let go of habitual reactions and assumptions as to how to respond; let go of expectations and our desire for certainty. We'll need to see things with fresh eyes, rather than those of the past, so that we can respond creatively to the challenges we face.

And, like the trees, we will need greater stability. We'll need to be stably anchored in ourselves, so that, when the unexpected suddenly arrives, we can remain relatively cool, calm, and collected; not be repeatedly thrown into fear and panic. If we lose our inner equanimity and react impulsively to every new development, we will become increasingly stressed and more prone to burnout. Nor will we be able to think so clearly as how best to respond.

How can we develop these qualities? This is where the psychological development mentioned earlier can play a valuable role. Needless to say, there are many schools of thought on this, and equally many paths, from psychotherapy and counseling to meditation and spiritual practices. Again, I won't survey them here. We are all at different stages in our journey with differing needs and backgrounds. What is important is that we each find ways to let go of whatever stands in the way of our being smarter, more creative, more resourceful, more compassionate human beings, responding to change with greater clarity and wisdom.

Now is the time we most need this psycho-spiritual awakening. Not to save ourselves from crises, but to help us navigate the times ahead with greater resilience.

A third factor that helps trees withstand a storm is being in a forest of trees. They soften the wind for each other. Similarly, we will need the support and companionship of others to soften the impact of the unexpected.

We will all feel vulnerable at times, needing to express our concerns and anxieties, and ask for emotional support. There will be the stress of adapting to new circumstances; emotional pain as we are forced to let go of cherished lifestyles and adjust to new realities; along with sadness and grief at some of the more distressing developments. And there may be times when we need to give material support, providing basics such as food, water, shelter, and healthcare to each other.

More than ever, we will need compassion and forgiveness. Forgiveness, not just of others, but of humanity itself. True forgiveness comes from understanding. If I put myself in another's shoes, I often can understand how they might have seen things and why they acted as they did. When I do, my judgements and grievances lose their grip and my anger begins to subside. Similarly with the world in which we now find ourselves. If we can understand how we got here—as I have tried to do in this book—then we can be more forgiving of ourselves. And, hopefully, accept this is how it is to run up against the inevitable consequences of ever-accelerating development.

For me, acceptance of the situation has brought with it some surprising shifts. I am not so angry at the groups whose views and actions I disagree with. I am no longer so shaken by the latest political shenanigans, economic swings, or social unrest. I can be more understanding, more forgiving, and, hopefully, more compassionate.

But it does not mean I no longer care for the world around me. I still want to do what I can to preserve the planet. But now I want to do so for the planet's own sake, rather than saving it for human beings. Perhaps the best we can do with our remaining years is to make sure we leave the Earth in as good a state as possible for the species that remain and those that will follow.

A New Story

It also leads me to a new story of humankind—not one of the optimistic "new stories" that tell us that if we just get our act together, free ourselves from outdated thinking, wake up to our true nature, and work together in harmony, we will be able to avert catastrophe and move on to a sustainable long-term future—but a new story of our place in the cosmos. And, for me, a much more magnificent one at that.

There are about 100 billion stars in an average galaxy, and an estimated ten trillion galaxies in the visible universe (which may be just a small fraction of the total universe, or universes). So that's around a sextillion stars. A thousand times more stars than the quintillion grains of sand in all Earth's beaches and deserts!

Of these stars it is thought that about one in twenty could have planets that are potentially habitable. How many of these do support life is harder to estimate. Even if life gets started on only one in a thousand—and that is a very conservative estimate—that's still hundreds of quadrillions of planets with life. How many would progress from simple cells to more complex cells and thence to multicellular life? Biologists think it could well have taken a lucky break for life on Earth to cross that threshold. But even putting the chances of that at another one-in-a-thousand, there could still be a 100 trillion planets with multicellular life.

On some of these, a rich diversity of species would emerge, and as they grew more complex would develop senses and nervous systems of some kind. From time to time, one of those species takes the step into tools and speech. A bud of creative intelligence appears.

On our planet it was preceded by billions of years of cellular evolution. Then by hundreds of millions of years of vertebrate evolution. And then, in just a few million years, our tool-using ancestors with larger brains appeared. With the advent of speech, the bud grew rapidly. Within a short time, cosmically speaking, it started to bloom, bursting into an exotic, multifaceted, cultural flowering. Billions of self-aware petals, seeking to become all they could be; to know all there is to know.

Within the time remaining we may still come to a full knowing of the world, both around us and within us. This does not mean knowing everything it is possible to know, but everything this particular intelligence could know, in this biological form, from this point in the universe.

Another bud of consciousness will have blossomed.

Exiting with grace

Here we are, wondrous beings, with unique gifts and abilities. We are capable of love and compassion, an appreciation of beauty, the creation of great art, music, and poetry. We are aware of our history, and how we came to be here. We have studied the world around us, and been awed by what we've discovered. We find meaning in our lives, a sense of justice, and an inner wisdom.

There is much to celebrate about us. The question is: Can we celebrate all that we are, while accepting that our species may be here for but a brief flash in cosmic time?

I am reminded of the so-called century plant that flowers once in twenty years or so. When it does finally bloom, we marvel at the giant stalk, holding high a magnificent array of flower-laden branches. The spectacle is made all the more awesome by the knowledge that it flowers but once; then dies, its purpose complete. Can we celebrate ourselves in a similar light? Another blossoming in the cosmos. An exquisitely beautiful flowering of consciousness. A miracle of creation.

Can we let go of the cherished belief that we are here to stay, rejoice in our existence, and live our final days with grace?

Despite knowing the journey, and where it leads,
I embrace it and welcome every moment.

~ Louise Banks in *Arrival*

Peter Russell, M.A., D.C.S.

Peter Russell studied mathematics and theoretical physics at Cambridge University (UK). Then, as he became increasingly fascinated by the mysteries of the human mind he changed to experimental psychology. Pursuing this interest, he traveled to India to study meditation and eastern philosophy, and on his return took up the first research post ever offered in Britain on the psychology of meditation. He also has a postgraduate degree in computer science, and conducted some of the early work on 3-dimensional displays, presaging by some thirty years the advent of virtual reality. In the 1970s, he pioneered the introduction of personal growth programs to corporations, running courses for senior management on meditation, creativity, stress management and sustainable development.

He coined the term "global brain" with his 1980s' bestseller of the same name in which he predicted the Internet and the impact it would have on humanity. He is the author of ten other books, including *The Brian Book*, *The Upanishads*, *Waking Up in Time*, *From Science to God*, and his most recent, *Letting Go of Nothing*.

He has been fascinated by the nature of exponential growth since he first came across it in his study of mathematics. In this book he shows how central this growth is to the future of humanity and to the challenges we face, revealing a new story of our place in the cosmos.

Made in the USA
Monee, IL
22 April 2023

32234007R00056